Process Development Symposium Europe 2023

Frankfurt am Main, Germany
26-28 June 2023

ISBN: 978-1-7138-8678-5

Printed from e-media with permission by:

Curran Associates, Inc.
57 Morehouse Lane
Red Hook, NY 12571

AIChE
120 Wall Street, FL 23
New York, NY 10005-4020

Phone: (800) 242-4363
Fax: (203) 775-5177

www.aiche.org

Additional copies of this publication are available from:

Curran Associates, Inc.
57 Morehouse Lane
Red Hook, NY 12571 USA
Phone: 845-758-0400
Fax: 845-758-2633
Email: curran@proceedings.com
Web: www.proceedings.com

TABLE OF CONTENTS

CONTINUOUS MANUFACTURING

PRODUCT DESIGN AND ENGINEERING

ENERGY TRANSITION

PROCESS DEVELOPMENT

BIOPROCESSING AND SYSTEM BIOLOGY

TOOLS & APPLICATIONS

2023 PDS EUROPE POSTER SESSION

Modern catalyst designs - the microchips of chemical processes

Authors

Perregaard, J. - Presenter, BASF

The presentation provides examples of the impact on technology by the market drive towards a new energy reality. Focus is on a sustainable future for both energy and chemical processes and BASF's own transition is used to illustrate the trends. In addition, the presentation shows how a modern approach to catalyst design and catalyst production enables the transition of these industries by allowing process design to make it's impact on the future operation of chemical plants, and new energy vectors will be addressed via concrete examples.

Digitalisation tools for efficient development of novel products and their robust manufacturing processes: development and democratisation

Authors

Birmingham, S. - Presenter, Siemens

In this talk we will provide an overview of modelling tools and use cases covering product and process development at different length scales, viz solvent selection, VLE predictions, in-vivo absorption, characterising kinetics (reaction, crystallization, drying. etc.), assessing impact of hydrodynamics on those kinetics and ultimately product quality and other process KPIs. We will also show examples where the digital twin that supported product and process development is used for operator training, soft sensing, nonlinear model predictive control, etc.

A New Tool Using Flow Chemistry Techniques for Continuous Crystallization of Pharmaceutical Drugs

Authors

Lecomte-Norrant, E. - Presenter, IPSOMEDIC SAS

In order to answer to the need of patients, it is important to develop innovative solutions to produce new drugs with the right particle size and polymorph in order to control the absorption of the drug by patient.

Then it could be a nice solution to propose a personalized medication, that means to prepare the tablet with the right dose with 3D printing for each patient in order to avoid an overconsumption of drug which can induce second effect for patient.

The new IPSOMEDICâ€™s lab installed in Aubagne (South in France), proposes the high quality right drug process with the right dose customized for each patient at low cost and safe conditions. The main objective of IPSOMEDIC & IPSOMEL laboratory, is to develop innovative continuous manufacturing processes of chemical and biological API beneficial to the patient.

The industrial unit will encompass flexible, safe, green and mobile modular units on skid . Production capacity can be easily expanded by doing the numbering up â€“ Technology to transfer this unit to other countries. This mode of construction and the implementation of new technologies is the only way to make the relocated production profitable.

The Aubagne Lab fully automated, with a special effort in data analysis, implements innovative reactions , separation modules ; Numbering up techniques
reduces the risk associated with the old scaling up methods.

Furthermore, Analytical techniques, such Raman,.. are key technology to control the particle size of crystallization step and also API concentration in each tablet.

Continuous Manufacture in the Broader Context of Sustainable Supply of Medicines

Authors

Rutter, A. - Presenter

Flow Chemistry, Continuous Manufacture has been divisive topic in Industry; some believe it is the future, others a distraction. I think Continuous Manufacture only makes sense when viewed as part of some broader socio-economics changes â€" the need to supply medicines reliably, affordably and without creating harm (environment/safety). In this talk I will use my journey starting with Petrochemical Manufacture to building some of the first pharmaceutical multistage API processes to illustrate why Continuous Manufacture matters â€" but only as part of some broader changes: namely automation, modularisation and digitalisation that support changing the current supply regime.

Biography

Andrew Rutter is an independent consultant with expertise in the application of Advanced Manufacturing Technologies, particularly in Medicine Manufacture. A Chemical Engineer by training he spent 20 years developing new processes API for GSK, including appointment, as EFPIA representative, to the Expert Working Group for ICH Q13. Prior to this he worked on Large Scale Petrochemical processes, and Process Economics for IBM. He currently advises many major Pharma companies, and Government organisations.

Foundations for Agile, Sustainable manufacturing

Authors

Alford, G. - Presenter, AstraZeneca

The current manufacturing infrastructure makes it difficult to offer the lead times and minimum order quantities that are required from the industry, ultimately leading to: investment at risk / or new product programme delay, high levels of inventory, low utilisation of manufacturing assets and potentially even stock-out or lost product sales opportunities.

We will look to understand what a factory would look like when the following themes are brought together, and to understand some of the common challenges that needs to be overcome to bring this vision to life

Key Themes that will be looked at include

- Modularity
- Modelling
- Equipment Characterisation
- Flexible Equipment Assets / Re-Tooling existing assets
- Decision Making Tools / Business Case
- Innovative SHE Assessment Tools
- Innovative Validation / Qualification Tools
- Digitalisation of human interactions

Screening proposal to take into account acceptable new substance

Authors

Lemaire, P. - Presenter, TotalEnergies

Lombard, A. - Presenter, DAIKIN CHEMICAL

Most of industry, regulators, politic and media are referring to sustainability, according from where you are coming from (academic, scientist, political, media, industry, no governmental organization,...) and where is your target (carbon print or environmental toxicology,...), everyone have a different position taking into account that this very widespread sustainability wording is not well defined.

For some it is only a question of carbon footprint at geological scale completely forgetting the today immediate effect on the substance on human health and environment. In this proposal we are going to focus more on the environmental toxicology screening hazard strategy to consider to lead to new minimize hazardous substance.

As we are talking screening which mean rapid and low cost testing, the strategy proposed is starting from the Persistency and Bioaccumulable and Toxic (PBT, human toxicology being part of the â€œTâ€) which is one of the main reason not going further in development of a new substance. Indeed today having a substance fulfilling anyone of those criteria can not be any more place to the generic market (but still on specific one according to risk assessment.

Product stewardship: understand risk assessment of new substance

Authors

Lemaire, P. - Presenter, TotalEnergies

Remonnay, I. - Presenter, Veolia

Lombard, A. - Presenter, DAIKIN CHEMICAL

Outcome of a screening hazard evaluation could lead to open widespread use if not major impact has been observed or restricted use in case of some issues has been observed. Indeed taking into account different regulations including REACH (Registration Evaluation Authorization of Chemical), PPP (Plant Protection Product), cosmetic or OSPARcom for instance, some hazard could be accepted or handle in a way of no more risk could remain.

According to business where the substance should be use, product steward should consider CLP/GHS regulation and found the adequation between substance concentration and substance action needed targeted to lead to an acceptable hazardous classification for the business. Indeed, there is difference between developing a new no hazardous substance and a new product (voluntary mixture of substances).

Knowing the screened final hazard evaluation of the substance or the product, either the substance and product can be used as such in any market or some more precaution should be taken into account to develop the commercial product with restriction of use or adapted precautionary used. A theoretical risk assessment evaluation should be done with the different use scenarii intended and acceptable used decided.

When green light is there the next step is to go to fulfill registration required according to the kind of use and generate more testing on toxicology and ecotoxicology taking into account environmental fate having a clear strategy based on tonnage band.

The Megatrends that re(shaping) the industry

Authors

Remonnay, I. - Presenter, Veolia

Industry 5.0 and the product of the future will be the result of the acceleration of the different transitions of industry 4.0, the current processes, products and people. The transition pathways will not be only a digital or technological shift but a global transformation of the mindset of people, policymakers and industry to embrace more broadly the components of the different transitions ongoing and linked to the megatrends for 2050, 2100 and beyond... and especially the scarcity of resources (skills, energy, raw materials, water...). Beyond the word â€œEcological Transitionâ€�, it means more than only climate change!

The definition of a strategy roadmap for delivering sustainability values and building sustainable industry 5.0 and products, with different stakeholders and ecosystems, need to identify the potential contributions to sustainable development and the key megatrends that will reshape the industry : the future of industry demography, the future of the work, the new models of governance, geopolitics, new economics as the Green DealS, the radical or existing innovation, the circular economy, digital technologies as artificial intelligence (AI) and robotics..., and the scarcity of the ressources (skill, raw material, energy, water, funding...). The industry should be innovative by gathering the existing and new blocks to build Industry 5.0 transition pathways by making industry, products, processes respect the planetary boundaries.

â€œIndustry 5.0 should not be understood as a chronological continuation of, nor an alternative to, the existing Industry 4.0 paradigm. It is the result of a forward-looking exercise to help frame how European industry and emerging societal trends and needs can co-exist.â€� European Commission, Directorate-General for Research and Innovation, Breque, M., De Nul, L., Petridis, A., Industry 5.0 : towards a sustainable, human-centric and resilient European industry, Publications Office, 2021,

Transitioning from Batch to Continuous API Crystallization to Establish a More Scalable Means to Kinetically Control the Purging of a Co-Precipitating Impurity

Authors

Glennon, B. - Presenter, University College Dublin

Morris, G., APC Ltd.

Keogh, A. P., APC Ltd.

Farid, U., APC Ltd.

Stumpf, A., Genentech, Inc.

In this presentation, APC discuss the application of continuous crystallization to control drug substance purity and crystal form in the isolation of Genentechâ€™s JAK1 inhibitor, GDC-4379 (Morris et al., 2022). Motivated by challenges encountered with the reliable purging of a regioisomer impurity in production-scale batch recrystallizations of crude GDC-4379, the adaptation of this impurity purging recrystallization to a continuous mixed suspension, mixed product removal (MSMPR) crystallizer enabled superior kinetic rejection of the regioisomer (97.9%) in comparison to the batch operation (32.4%). Furthermore, by operating the MSMPR crystallizer at sufficient water content and temperature, the ability to selectively crystallize the required hydrate form A of GDC-4379 for the active pharmaceutical ingredient (API) directly from the solvent system of the impurity purging recrystallization (DMSO/MeOH) was confirmed. The results of the development effort provided proof of concept that a continuous crystallization process could achieve the regioisomer purge and API form control requirements of the GDC-4379 drug substance and provide a more scalable and streamlined alternative to the two-step batch recrystallization process used in production.

References

Morris, G., Keogh, A.P., Farid, U., Stumpf, A., 2022. Development of an impurity and hydrate form controlling continuous crystallization to telescope a two-step batch recrystallization in the GDC-4379 drug substance process. Chem. Eng. Res. Des. 183, 608â€"622. https://doi.org/10.1016/j.cherd.2022.02.022

Biography

Brian Glennon is a Professor of Chemical Engineering in University College Dublin (UCD) and co-founder of APC and VLE Therapeutics. He holds a BE and PhD in Chemical Engineering from UCD. Prior to joining UCD, he worked with Merck in Ireland. Since joining UCD in 1995, his research interests have focused on pharmaceutical process engineering, with particular emphasis on crystallization, bioprocess design and continuous manufacturing. In his role as Chief Technology Officer in APC and VLE, he leads the internal research programmes and is charged with driving innovation across both organisations.

Process Intensification Flow Reactors: A Well Accepted Technology through Various Chemistry Fields

Authors

Gauron, G. - Presenter

Winter, M. - Presenter, Corning SAS

Transferring chemical synthesis from traditional batch technology to continuous flow methodologies brings significant advantages when looking for a rapid process scale up. In the pharmaceuticals and fine and specialty chemicals industries product quality and availability have become todayâ€™s crucial parameters and it is often not possible to maintain optimum product quality when scaling up a process in a short period of time using traditional approaches. A seamless move from lab-scale to larger scale and production volumes using the same technology platform is particularly attractive in the areas of cost reduction, safety management and response times.

Problems associated with batch process scaling can be overcome via straight-forward methodologies utilizing a consistent performance continuous process platform which will be described: 1000x improvement in heat transfer, 10-100x enhancements in multiphase mixing, x/1000 reduction in chemical holdup comparing with conventional stirred batch reactors. This seamless scaling also reduces development and production costs leading to a drastically reduced time to market.

This talk will focus on several cases of applications that have been seamlessly transferred from lab scale directly to industrial production using continuous flow and the benefit for the industry will be highlighted. Producers coming from Agrochemicals, Pharmaceuticals, Energetic materials, Fine and Specialty chemicals, Nano materials are already using such technology up to more than 10.000mT/year production.

Let there be bright light

Authors

Metten, B. - Presenter, Ajinomoto Bio-Pharma

The scale up of both homogenous or heterogenous photochemical processes is far from trivial. The reactor selection is one side of the story, but evenly important is the light source and its characterization. A process development on a homogeneous example will be disclosed and compared to other pilot tests, demonstrating the potential of further scale up. Examples of heterogeneous photochemical processes will be presented to show the first innovative proof of concepts.

Decentralised Production Realised using Continuous Manufacturing - Harnessing the Power of Flow Chemistry from Lab to Production

Authors

Wiles, C. - Presenter, University of Hull

Small-footprint, flexible, modular reaction units bring an opportunity to disrupt conventional supply chains via strategies such as point of use production & just in time manufacturing. In addition, these small, turn-key units lend themselves to replication to scale output & since the location of such units does not need to be fixed, continuous manufacturing enables rapid deployment of manufacturing capabilities to new locations â€" particularly to those without conventional chemical manufacturing infrastructure. Case studies will be used to highlight the replication of existing continuous manufacturing units, enabling the deployment of additional production capability in the UK, USA, France & Belgium.

Combining Surrogate Modeling, Superstructures and Neural-Aided Optimization for Design of Innovative Intensified Processes

Authors

Commenge, J. M. - Presenter

Pina-Martinez, A. D., Université de Lorraine

Wu, D., Université de Lorraine

Behloul, C. R., Université de Lorraine

Falk, L., University of Lorraine - CNRS

The energy transition requires certain chemical processes to be redesigned to adapt to sustainable operation (biobased inputs, intermittency, decentralized production, etc.). These changes necessitate a reshape of the process structure. Concurrently, intensified multifunctional units make it possible to take advantage of the synergies between transformation functions (reactions, transfers, separations, etc.) within a single piece of equipment. Including these units in a process may lead to a simplified process. Therefore, design of energy-efficient processes necessitates to evaluate new process structures.

Process synthesis methods have considerably evolved from heuristics-based to mathematical-programming based optimization (MINLP) and evolutionary approaches. Numerous tools are available to widen the scope of attainable structures. Implementation of superstructures in CAPD software is possible through conditional splitters (switches) and optimization libraries. Extensions enable to combine process-engineering criteria and product-engineering formulations. User-defined functions may consider fast surrogate models, either to accelerate optimization or to substitute time-consuming models. Multi-objective optimization with constraints are helped by neural networks to accelerate convergence in the range of decision variables.

This work illustrates these opportunities with examples. Efficiency of micro-CHP systems is increased by compact heat-exchanger reactors: constrained multi-objective optimization is slowed down by the simulation of these units. Iterative fitting of a neural-based surrogate model of the objectives and constraints enables to explore the optimality domain and focus within the area of interest. Combined heat and power thermodynamic cycles can be taylor-designed by merging product and process design approaches: several fluids are screened while the optimal flowsheet is determined by superstructure approach including bleeding, reheating or regeneration as structural alternatives. For the synthesis of Dimethyl Ether, multifunctional reactors (coupling reaction, heat-exchange and separation) are relevant to reach very high yields, but their integration in a CAPE environment is delicate: Physics-Informed Neural Networks (PINNs) fasten the optimization step for further comparison with conventional technologies.

Process Synthesis and Machine Learning for CO2 / H2 Membrane Separation

Authors

Piccialli, V. - Presenter, University of Rome Tor Vergata

Castel, C., Laboratoire Réaction et Génie des Procédés LRGP- CNRS, Université de Lorraine

Addis, B., Université de Lorraine

Misener, R.

Macali, A., University of Lorraine

We propose a system design procedure for membrane gas separation based on a superstructure-based optimization strategy. We define a mathematical programming model of the system representing the superstructure. Instead of using the â€œclassicalâ€? discretization of differential equations, we represent the membrane functioning using a surrogate model built by training a neural network. The resulting optimization problem is a non-convex non-linear programming problem, which is solved by a global optimization technique. We consider hydrogen purification from a pre-combustion gas mixture as a case study.

We model two different commercially available membranes (UBE BH and Polaris from MTR) building a machine model for each one of them based on a data set obtained using simulation. Training an accurate model for each membrane type can require a large set of data points and hence a high computational time. To overcome this issue, we propose a data augmentation technique that starts from a small-size data set and improves it sequentially by adding points. Instead of using standard augmentation techniques, we devised a strategy that is based on the optimization itself which ensures an accurate model in the regions of interest for the case study.

The results obtained for H2 purity of 90 and 99%, with different recovery requirements are detailed and discussed. A clear advantage emerges in using the surrogate model since it makes the optimization process numerically more stable, and much faster. All the solutions produced are validated and refined by a simulation approach.

Process Synthesis Using Commercial Tools: Feedback from R&D to Industry for Electric Production Applications

Authors

Neveux, T. - Presenter, EDF R&D

Mecheri, M., EDF Therm Engineering department

Le Moullec, Y., EDF R&D

Systematic process synthesis methods and tools have been developed for decades together with the advance in computing, with profusive academic activities but with less reports from industry on the use of these approaches.

This study intends providing feedback on superstructure multi-objective optimization experiments conducted together by the EDF R&D and engineering divisions, trying to shift from research approach towards use in real engineering projects.

To do so, the methodology is based on the coupling between a commercial existing process simulation software and an external solver. The process simulator (ProSimPlus) is where the synthesis problem and the exploration space are defined, leading to the definition of a superstructure. The external solver (e.g., MIDACO for multi-objective optimization in this study) is used to solve the underlying mathematical problem, seeking for optimal structural alternatives (flowsheet) and associated degrees of freedom (design and operating parameters) within the given constraints.

So far, studied cases cover various types of industrial processes (e.g., power cycles, separation processes) and stages of industrial projects (technology benchmark, conceptual study, assisting of process design). Experiences will first be presented in the design phases of an industrial project (with examples of power cycle design, such as innovative supercritical CO_2 cycles). Other examples cover usages for technology benchmarks, using multi-objective optimization to provide insights to the process designer on the physical possibilities of each process (with example for the comparison of membrane technologies for the water purification).

Current perspectives can be drawn from these first results, in relation to ongoing challenges in the energy sector with a new landscape for mature and innovative technologies, and to new prospective tools such as the rise of generative approaches for process design.

Improvement of Plant's Reliability through Web-Based Online Monitoring (Digital Manufacturing Excellence)

Authors

Qosim, H. - Presenter, Pupuk Indonesia

Baqir, M., PT Pupuk Indonesia

Atminanto, A., PT Pupuk Indonesia (Persero)

Gustaman, G. G., PT Pupuk Indonesia (Persero)

Saikhu, M., PT Pupuk Indonesia (Persero)

PT Pupuk Indonesia (Persero) has 14 total of Ammonia & Urea Plants across Indonesia. Our production facilities are managed by our 5 subsidiaries with total capacity of Ammonia **up to 7 million ton per year and up to 9,4 million ton yearly.** That big production capacity and high value of asset, it is essential to maintain reliability at high level. High output and reliability of the Plants are crucial because Gas form up to 70% cost of production. If Normal operation of the Plant is disturbed or even cause the plant to be shut down, Cost of Production will go up. The effect on Environment is as much as effect on cost of production. Not to mention global competition in international market, we require a tool to monitor process production thoroughly to keep production high and reliable

Board of Management in production facility and at Holding office, Pupuk Indonesia faces challenges how to monitor and evaluate to prevent shutdown of the Plants. Our team, collaboration among Production department at holding, subsidiaries and IT succeeded to create innovation in order to monitor process production online via website. We call it as Digital Manufacturing Excellence (DMEx).

DMEx is a web based system contains real time operational data send directly from Historian in Distributed Control System in production facilities in our Subsidiaries. This Data then used to monitor process operation real time. It is also utilized to address instability in operation parameter. History of the operational data readily used to address root causes if the breakdown occurs. Furthermore, instable parameter of operation is used as precaution for Plantsâ€™ operator to adjust accordingly

This Digitalization has been assisting board of plantâ€™s management from supervisor to Plant manager to monitor plantsâ€™ operational and helping board of management in Holding company to take on strategic decisions.

An Orchestrated Laboratory using Artificial Intelligence to Accelerate Innovation in Chemistry

Authors

Company, R. - Presenter, Solvay

Artificial intelligence is being a revolution in technology for daily life and technical applications. Nowadays algorithms can predict consumers' habits and even answer questions with incredible accuracy. The impact of this technology is also observed on research and a developpement of chemical synthesis and other applications on the chemistry field. Particularly in the case of chemical formulation these algorithms can be used to accelerate and reduce the number of experiences necessary to explore formulation phase diagrams. Design of experiments combined with Bayesian optimization could help to explore and understand multivariable space which can be extremely complex to understand only using the human brain. In this particular case we use a robotic platform to perform the experiments, combined with artificial intelligence algorithms that will predict the experience to perform and will feed and optimize the prediction models based on the results obtained. The case that we choose for the study is a formulation for Home and Personal Care, based on surfactants, polymers and salts to optimize viscosity of the shampoos. The complexity of the interactions between the components from the formulation needs to be coupled with the scientific knowledge to ensure the optimization with good accuracy. It means that artificial intelligence will not be able to replace humans but is definitely a tool to accelerate research in a complementary interaction.

Designing Waste Plastics Recycling Processes Based on Machine-Learning Thermodynamics Predictions

Authors

Bosetti, L. - Presenter, ETH Zurich

Schilling, J., ETH Zurich

Winter, B. A., ETH Zurich

Lindfeld, J., ETH Zurich

Bardow, A., RWTH Aachen University

The chemical industry's shift towards a net-zero CO2 emissions future requires a circular economy where both mechanical and chemical recycling play a key role in reducing plastic pollution and substituting fossil-based feedstocks.[1] However, chemical recycling of Polyurethane (PUR) rigid foams, which leads to complex mixtures, often requires high-energy separation processes, undermining the overall sustainability of the recycling process.

To address this challenge, we aim to develop low-energy separation pathways for the chemical recycling of PUR rigid foams. The end-of-life PUR is processed in a catalytic process resulting in a mixture of aniline, propylene glycol, and other by-products. Distillation columns are commonly used to recover aniline, but the energy demand and the total cost of the separation are expected to be high. To address this, extraction could potentially be integrated into the process design, but identifying a suitable solvent is a challenge.

For this purpose, precise methods are required to calculate thermodynamic properties. Recently, we showed that a machine-learning approach (SPT) can precisely predict activity coefficients for a broad range of molecules.[2] Here, we expand the scope of SPT to determine the solvents' thermodynamic properties. Based on the SPT predictions, we can screen a large database of solvents suitable for aniline recovery. The resulting method yields a ranked list of solvents potentially suitable for the chemical recycling of PUR.

Our work demonstrates the conceptual design of the separation of a reaction mixture coming from PUR foam recycling enabled by machine learning. This resulting design method allows screening rapidly and accurately a very large number of solvents for which the thermodynamic properties are not experimentally available. The computational efficiency method renders the method particularly suited for future integration into the framework for integrated computer-aided molecular and process design (CAMPD).

1. Meys, et al., *Science* 2021. DOI: 10.1126/science.abg9853
2. Winter, et al., 2022. https://doi.org/10.48550/arXiv.2209.04135

The Use of Mathematical Models and Digital Twins to Improve Scale up. What Is the Role of a Systematic Scale Down for the Experimental Design of Lab Experiments for Smart Scale up?

Authors

Levis, M. - Presenter

For direct scale up of a chemical process it is important to master the performance as well of involved chemical reactions as of the used manufacturing equipment.. Comprehensive understanding how a chemical process will behave in production-scale equipment is the key for success to make predictions for process safety, quality and economics in scale-up. Various mathematical models, which describe reaction kinetics, unit operations, mixing properties assist us in process development.

Based on the case study of a complex heterogeneous hydrogenation, it is described, how mathematical models were used in our scale down / scale up approach. Digital twins of manufacturing and lab reactors were established to design laboratory experiments - generating meaningful data for direct scale up from lab to plant. A new concept to ensure comparable heat transfer and geometrical equivalence to lab experiments was introduced. 3D printing of stirrers and baffles is used to vary and study the effect of hydrogen transfer, mixing properties and dispersion of solids. A new concept for heat transfer enables us to employ identical rate of heating surface to reaction volume over various scales. Heat transfer rates as well as surface temperature can be adjusted to mimic the targeted equipment. A special designed colling finger and a new cryostat proptotype hat to be created.

Experimental design, mathematical models for kinetics, mass transfer and heat transfer are combined in a "tool box". A scale down reactor (lab scale twin) was manufactured to validate the model and to test optimized process parameters. Based on the lab experiments the chemical process parameters were optimized. Results and predictions were succesfully compared with data from manufacturing scale and for further refinement of the model.

Publications: https://doi.org/10.2533/chimia.2021.948; OPRD publication planned

Process Optimization across the Water-Energy Nexus

Authors

Duval, Q. - Presenter, ProSim SA

Turzo, G., ProSim SA

Baudouin, O., ProSim

Dechelotte, S., ProSim SA

To give an order of magnitude, food production currently accounts for 90% of global consumption of fresh water and 30% of global energy consumption and much of this water and energy is wasted [1]. Moreover, a common feature of industrial processes is their use of fossil fuels as the primary source of energy, where a large part of that energy consumption is spent on the production of utilities. The scientific world makes a significant effort to find alternative sources of energy. However, even by the most optimistic assessments, these alternatives are long-term solutions and many projections show that in the near future, fossil fuels will remain the primary sources of energy, in particular for the process industries. So practical process optimization tools are critical for chemical engineers to understand the connection between water and energy, and then to eliminate both water and energy waste in their plants.

Pinch analysis [2] is a well-known methodology to optimize process energy consumption. It allows the determination of the process minimum utilities requirement. *Simulis Pinch*, software used directly in MicrosoftÂ® Excel, has been developed to perform energy integration calculations with the pinch method, while also systematically considering real-world engineering constraints. As part of the MINIMEAU project funded by the French ANR [3], *Simulis Pinch* has been extended to the Water Pinch Analysis (WPA) and the design of efficient water networks. It has been successfully applied for the optimization of water resources of various processes in the food industry. Used in connection with *ProSimPlus*, steady state processes simulation software, *Simulis Pinch* allows to optimize the processes across the water-energy nexus.

[1] Izabela Balicka, â€œThe Food-Energy-Water Nexus: A Complex Balanceâ€?, CEP, 2020

[2] Bodo Linnhoff, PhD thesis â€œThermodynamic Analysis in the Design of Process Networksâ€?, Leeds University, 1979

[3] https://minimeau.fr/

Development and Demonstration of an Ultra-High-Temperature Continuous Racemization Process

Authors

Sheeran, J. - Presenter

Campbell, K., Snapdragon Chemistry

Fang, Y. Q., Snapdragon Chemistry

Nagy, K., Snapdragon Chemistry

Ford, D. D., Snapdragon Chemistry

Herein we describe a continuous flow process for the thermal racemization of an undesired atropisomer to facilitate recovery of the desired form from the racemic mixture. A kinetic model was developed with the ability to predict the degree of racemization at varying temperatures. The high activation energy for the process required it to be carried out over 300°C under continuous flow conditions using a superheated solvent stream. Process safety was a key factor in the reactor design due to the operating conditions and a potential thermal event observed during the characterization work. The process was demonstrated in a lab-scale (~100 mL) tubular reactor before being scaled up to a pilot-scale (~1 L) reactor that processed >15 L of feed in a day. Following the reaction, the product was recovered via cooling crystallization as a high purity racemic material suitable for further processing.

Engineering of the Future Is a Key Enabler to Deliver Fast and Robust Manufacturing Plants

Authors

Sakly, M. - Presenter

This conference will focus on the methodology to move from document management to data management from preliminary through construction to delivery of the project . The executive summary of the presentation is

- all changes during PID s management to minimize the redo and get a robust design.

- allow a continuity during the construction phase using all relevant tools in order to avoid dichotomy between design and construction

- deliver a physical asset with a digital twin facility

This conference will introduce all the key aspects such as modualrity in the design , modualrity in the construction in addition to the key success factors of interoperability , flexibility of the design

This conference will emphasize focus on modularity of design , modularity of construction, modularity during a startup in order to accompain the business needs basically when we talk about the launching of new products.

This modularity should be designed taking into account the automation strategy

Robots and automatization in clean room bioprocessing : the solution towards GMP Annex 1 securing bioproducts quality and increasing productivity and sustainability

Authors

de Malliard, H. - Presenter, MGA Technologies

R&D in biotechnologies for all medical applications and bioprocessing, new advanced therapy medicinal products are proposing candidate innovative products with jeopardized future industrial production potential. This uncertainty leads to high financial risk in building industrial facilities at early stage of clinical trials and high costs for these new medicines. Hence the development of flexible and evolutive facilities capable of bioproduction for multiprocess and multiproducts is more than ever necessary. This session is technological driven, gathering technical bricks already available and foreseeing those easy to develop to achieve the vision of the facility of the futur fully automatized, robotized and digitalized.

Keywords: robotics, mechatronics, single-use bioprocessing, mechatronic sterile welders and sealers, MTP standards, 5G, augmented operators, AI, ML, digital twins, modular evolutive flexible multiproducts and multiprocess facilities, facility of the future

Digitalizing Manufacturing of Products

Authors

Frenzel, P. - Presenter, Covestro

Covestro aims for net zero scope 1 and 2 emissions by 2035. This transformation is based on three levers, green electricity, green steam, and more sustainable manufacturing processes. For this last one, aspects such as digitalization of conceptual design are essential. In this presentation we will share our vision of a full digitalized plant, with improvements in safety and efficiency with the help of automation and the use of data analytics, providing concrete examples of already existing applications. We also present a revised conceptual design framework, with the focus on circular economy projects as part of our journey towards becoming fully circular.

Next-generation sustainable bioproducts

Authors

Noorman, H. J. - Presenter, DSM

Intensified agriculture and fossil fuels have served mankind with a wide range of abundant and affordable food products, energy carriers and consumer goods. Today however, planetary boundaries are jeopardizing current living standards, the need to reinforce underdeveloped regions, and accommodating further growth until 2050. This calls for a next generation of products that can replace less sustainable practices, and also are innovative, providing better functions and sustainable solutions. Bioproducts, made from low-carbon emitting feedstocks and renewable energy with the help of biology, e.g. via fermentation, are expected to take a leading role in this transition. In this talk several state-of-the-art examples will be provided, and also it will be outlined what will likely be main future bioproducts, how they best can be manufactured on industrial scale, and what is needed to speed up and overcome the â€žinnovation valley of death.

Applying â€žzero-thinkingâ€œ will be key to success. How to maximize industrial performance with zero carbon and energy spills, zero use of fossil inputs, zero losses in development time and zero emissions to the environment?

Two most critical steps are value chain re-design, and robust design of the industrial fermentation. Feedstocks made from CO_2 using renewable energy will be complementary to land-derived feedstocks such as sugars, which are associated with land-use issues such as irrigation water use, biodiversity loss and rainforest destruction. Industrial bioprocesses should become more tolerant to industrial â€žstressesâ€œ such as concentration gradients and product inhibition, and bioreactors better supported by computational tools that can reproduce the process with high resolution and fidelity, enabling digital twins to allow best-informed decision-making in operations. Such simulation methods, however, today still require massive computation efforts and are not yet fit for broad, real-time deployment. Here, machine learning and artificial intelligence can be deployed to speed up and improve computation efficiency. Latest insights on these developents will be shared, all key to secure commercialization of next-generation bioproducts and the transition to a more sustainable society.

A new Laboratory using flow chemistry Techniques for continuous production of Pharmaceutical drugs– Case of PARACETAMOL

Authors

Lecomte-Norrant, E. - Presenter, IPSOMEDIC SAS

In order to answer to the need of patients, it is important to develop innovative solutions to produce new drugs with flexible manufacturing units. To be factual by using all disruptive innovations in different areas like intensification of process using continuous micro reactors controlled by PAT, Artificial Intelligence for retrosynthesis, Big data analysis, new materials of construction, microelectromechanical systems (MEMS), Internet of Things (IoT), 3D printing ...

This is the only way to relocate pharmaceuticals production which has been externalized to take advantage of local low manpower costs and regain our sovereignty

The new IPSOMEDICâ€™s lab installed in Aubagne (South in France), proposes the high quality right drug process with the right dose customized for each patient at low cost and safe conditions.

The main objective of IPSOMEDIC & IPSOMEL laboratory, is to develop innovative continuous manufacturing processes of chemical and biological API beneficial to the patient.

It has been developed a full continuous process for the production of Paracetamol (Acetominophen). This new process will be the heart of the IPSOPHENE project whose the plant will be located in Toulouse.

The industrial unit will encompass flexible, safe, green, and mobile modular units on skid. Production capacity can be easily expanded by doing the numbering up â€“ Technology to transfer this unit to other countries. This mode of construction and the implementation of new technologies is the only way to make the relocated production profitable.

The Aubagne Lab fully automated, with a special effort in data analysis, implements innovative reactions , separation modules ; Numbering up techniques reduces the risk associated with the old scaling up methods . Furthermore, Analytical techniques, such Raman, NIR, UV,.. are key technology to control the full process over time.

Digital design and operation of robust continuous drug substance manufacture processes: Industry cases & value statements

Authors

Bermingham, S. - Presenter, Siemens

Mahadevan, M., Process Systems Enterprise

In this presentation, we will describe applications demonstrating the use of mechanistic models for robust design of continuous drug substance manufacturing processes. This includes model calibration of relevant unit operations, development of dynamic integration flowsheet models and investigation of the impact of process disturbances and model uncertainties on Critical Quality Attributes (CQAs).

We will highlight approaches for characterizing residence time distributions and process dynamics to analyze disturbance rejection and material traceability (lot genealogy). We will present a novel modeling framework to track lot composition throughout a Continuous Drug Substance Manufacturing (CDSM) process, which can be useful in assessing regulatory quality and batch definition in production campaigns. Lastly, opportunities for mechanistic model-based digital operation will be described, including testing control strategies, soft sensing, operator training, and model predictive control.

ContiUnity® a True Modular Process Platform for Continuous Manufacturing

Authors

Bolt, W. - Presenter

In pharmaceutical and fine chemicals industry there is a continuing drive for multi-purpose production. With the smaller quantities, localized production and new continuous technologies new processing systems are required. With the technology of today a digitalized and alternative flexible use of process systems has come available, which are the enablers for realizing flexible and standardized hardware solutions to produce in the required multi-purpose environment.

At Zeton a modular concept is developed which includes process specific solutions as well as the flexible integration (and hardware solution) into existing production environments. This concept is also applied for customer specific processes, which includes a process orchestration approach, applying an overall design approach and use of standardized modules.

In this presentation the development route is presented, using various realized projects and internal development experiences. It will give insight on the requirements for this new modular concept to be successful in a fast developing industry.

FAST – Pfizer's journey into API continuous processing

Authors

Hardy, R. - Presenter, Pfizer

Flexible API Synthesis Technology, otherwise known as FAST, is Pfizerâ€™s Small Molecule Active Pharmaceutical (API) technology strategy. This joint initiative between Worldwide Research and Development (WRD) and Pfizer Global Supply (PGS), the manufacturing and supply division of the company, is delivering the tools needed to implement continuous manufacturing processes. The investment includes new and modified facilities, introduction of new laboratory equipment and provision of modular manufacturing equipment. Cross functional teams are developing guidance and updating existing workflows to support the development and introduction of continuous processes. This presentation will cover all these aspects and conclude with a case study.

Continuous processing and Digitalisation, the future of Pharma to supply medicine in a robust, agile and cost effective way to the patients

Authors

Susanne, F. - Presenter, GlaxoSmithKline

Continuous processing (CP) is a manufacturing methodology that dominates the production of commodity chemicals. CP is seen as the method of choice for delivering products associated with the oil and gas, food, and polymer industries s. In contrast, the pharmaceutical industry has only started to focus on the application of continuous processes. Two primary drivers have driven this renewed interest over the past decade. Firstly, flow chemistry enables a diversity of reactions from which to synthesize complex molecule. Secondly, supply chain and factory utilization benefits are beginning to be realized providing added-value benefits.

Sanofi is committed to the acceleration of availability of high-quality medicines for patients. Across the CMC small molecule organization, continuous manufacturing is being implemented to improve and accelerate the development and commercialization of processes delivering high degree of robustness, high eco-design potential and sustainability. Sanofi has adopted a *flow-by-first-intent* strategy which is transforming its current small molecule portfolio from one comprised of exclusively batch manufacturing processes into one based on hybrid batch-continuous processes.

Novel Process for Valuable By-Products Recovery from Hot-Liquid Water Pretreatment of Biomass

Authors

Jankovic, T. - Presenter, TU Delft

Straathof, A. J. J., Delft University of Technology

Kiss, A. A., Delft University of Technology

Hot-liquid water pretreatment of lignocellulosic biomass is an environmentally friendly method with relatively mild process conditions and low costs. However, the value of the liquid remaining after the pretreatment is usually underestimated. The novelty of this research is an enhancement of the biomass pretreatment step through recovery of valuable by-products (acetic acid, formic acid, furfural and 5-hydroxymethyl furfural (HMF)) from this liquid, which would otherwise be waste stream.

Different methods are reported for recovery of mentioned components from aqueous solutions, but yields are often insufficiently high, downstream processing is complex due to the usage of additional chemicals and research is done only on a lab-level, without scaling-up. Therefore, distillation is used as the main separation method in this work. The main challenge is highly diluted feed stream (>96wt% water) and thermodynamic constraints due to the formation of three azeotropes (water-furfural, water-acetic acid-formic acid and water-formic acid), together with the tangent pinch in water-acetic acid mixtures. Aspen Plus was effectively used to develop rigorous simulations for the full process operation.

About 78.8% of acetic acid from initial liquid is obtained as a high-purity product (99.8wt%), while the rest is recycled, together with most of the water, to the biomass pretreatment step. Recoveries of formic acid, furfural and HMF are higher than 99.5% and product purities satisfy market requirements (73.6wt% formic acid, 97.9wt% furfural and 100wt%HMF). Mechanical vapor recompression and heat integration are implemented to decrease energy use. This results in 88.0% reduction in primary energy requirements (from 75.13 to 8.99 kWh/$kg_{product}$), 76.6% reduction in total annual costs (from 2.41 to 0.56 \$/$kg_{product}$) and 99.7% reduction in CO_2 emission (from 10.84 to 0.03 kg_{CO2}/$kg_{product}$). In conclusion, the original contribution of this work is an optimal waste-to-value process which can enhance economic viability and competitiveness of biorefineries compared to fossil-based fuel production alternatives.

The laboratory of the future: Impact of AI NEW TRENDS

Authors

Lucas, V. - Presenter, CLIMALIFE

What are the impacts of the digital transition on chemical laboratories? What is the evolution of their state of mind regarding the new tools: BIG DATA, diffusion of AI. How and when to use AI and for what benefit?

The aim of this session is to give an overview of new trends in using the capabilities of digital tools to accelerate chemical synthesis or formulation of new products or to improve existing ones.

In my talk, based on my own experience with AI in formulation, I will highlight some of the strengths and weaknesses of this transition we are facing and illustrate with examples the area explored. I will then give the floor to the other speakers, experts in AI in chemistry, to illustrate the opportunities of AI to set up the laboratory of the future and to allow us to reach the objectives and to revisit chemistry in a new way.

Design of Particulate Products – Continuous Synthesis, Nanoparticle Chromatography and Multidimensional Characterization

Authors

Peukert, W. - Presenter, University of Erlangen-Nuremberg

Today, particle technology moves forward from careful control of â€œonlyâ€ particle size to systems with higher complexity, i.e. shape, surface, structure and composition, both on the level of single particles and particle ensembles. The applications define the required product properties, which cover both classical fields of process technology in the chemical industry as well as new emerging fields of electronics, energy and environmental technologies, life sciences, materials science, nanotechnology and photonic technologies. We highlight aspects of these trends by focusing on synthesis, property classification and multidimensional characterization of nanoparticle systems. These are integrated into well-defined structures of mesocrystals, supraparticles and thin functional films, all with interesting functional properties.

Based on unifying principles for the formation of particulate products, which are widely applicable to many different kinds of products including solid, liquid, and even gaseous particles, we discuss options for scalable production. Despite the amazing progress in lab-scale material synthesis, the transfer to industrial application is much less developed and is still a limiting challenge. By careful consideration of the driving forces for particle formation in batch and continuous systems, we show how scaling laws can be derived and applied to several material classes including noble metal, semiconductor and drug nanoparticles. Any process design is closely related to careful characterization of the evolving structures, at best applied directly in the process. Typically, synthesis is coupled to purification, fractionation and formulation. In particular, property classification is a major challenge. For the latter, we present recent results for size- and shape-selective nanoparticle chromatography. Examples of product design related to optical properties as studied in the collaborative research center â€œDesign of Particulate Productsâ€ will be highlighted.

Funding of CRC1411 (https://www.crc1411.research.fau.eu/) by DFG is gratefully acknowledged.

Effect of Gaseous Contaminants on Electrochemical CO2 Reduction to C2+ Products

Authors

Sajeev Kumar, A. - Presenter

Fu, S., Delft University of Technology

Ramdin, M., Delft University of Technology

de Jong, W., Delft University of Technology

Van Ommen, J. R., Delft University of Technology

Kortlever, R., Delft University of Technology

Electrochemical conversion of CO_2 into valuable fuels and chemicals has gained attention owing to the mild conditions applied and the possibility to integrate renewable energy sources to power the process. One of the major drawbacks of CO_2 point sources such as the chemical industry or power plants is the presence of contaminants such as SO_2, H_2S, COS and NO_x which could be detrimental to the copper catalysts that drive the conversion of CO_2 to C_{2+} products. The industrial feedstock cleaning procedures are typically energy and capital intensive, making it desirable to study the effect of various contaminants on the CO_2RR and to determine the limits of each contaminant in order to derive an optimal feedstock cleaning procedure.

In this work, we study the influence of various concentrations of sulfur-based gaseous contaminants such as SO_2, H_2S and COS in the CO_2 feed, on the selectivity, product distribution and catalyst stability during the CO_2RR. Presence of SO_2 above 100 ppm significantly suppresses the C_{2+} products and CO at -1.0 V vs RHE, however, at -1.2 V vs RHE, this is not observed until above 1000 ppm. The main products observed under the suppressed conditions are hydrogen and formate. With H_2S, a similar effect is observed even at 10 ppm at -1.0 V vs RHE, while at -1.2 V vs RHE, this is not observed until above 100 ppm. COS proved to be the most detrimental to copper, suppressing the C_{2+} products even at 10 ppm at -1.0 V and -1.2 V vs RHE. Recovery experiments show that it takes 1-4 hours of purging with pure CO_2 and a constant voltage supply to (partially) regenerate the copper catalyst after exposure to 1 h of 10000 ppm of $SO_2/H_2S/COS$, making the continuous operation of CO_2RR plant difficult in the presence of these contaminants.

Post-Combustion CO2 Capture By Absorption-Regeneration Process: How Reducing Its Energy Consumption ?

Authors

Thomas, D. - Presenter

The implementation of Carbon Capture, Utilization and/or Storage (CCUS) process chain appears as a necessity to reduce the CO_2 emissions from different industrial sectors, especially â€œprocess CO_2 emissionsâ€� such as for example in the cement, lime, glass and steel industries. Focusing on the most technologically mature carbon capture process, namely the post-combustion absorption-regeneration process using amine(s)-based solvents, one of the major obstacles for its implementation is still its cost and especially its very high energy consumption for the solvent regeneration step. For addressing this issue, different energy reduction pathways were investigated experimentally and/or through the development of Aspen Plus™ simulations. Upstream of the process, the possibility to increase the flue gas CO_2 content by applying partial oxy-combustion and/or flue gas recirculation was studied. Inside the process, more efficient and innovative mixtures of solvents such as demixing solutions were envisaged. Moreover, at the configurational level, advanced process configurations were simulated and compared on a techno-economic point of view. Among the results, it was shown that implementing an advanced process configuration (e.g. Inter-Cooling Absorber + Rich Vapor Compression + Rich Solvent Splitting and Preheating, with methyldiethanolamine (MDEA) + piperazine (PZ) as a activated solvent) or using a demixing process (e.g. composed of diethylethanolamine (DEEA) and methyl-amino-propylamine (MAPA)) are both very efficient pathways for reducing the thermal energy demand of the process (i.e. up to 40% savings in comparison with a conventional process using monoethanolamine (MEA)). In terms of CAPEX, the advantage of the demixing technology was highlighted as it allows to achieve very high energy savings but with a more limited CAPEX investment than with a more complex process configuration. The research for new demixing solvents, less volatile and degradable than DEEA+MAPA blends, but keeping similar energy performances, is still under progress.

Electricity only? Potential and limits to decarbonize primary energy supply in Europe highlighted by an eMethanol case study for Spain

Authors

König, S. - Presenter, Air Liquide

Europe has always been an importer of primary energy, importing oil and gas. But could we be self-sufficient by using renewable energy sources (RES) only? Or do we need energy imports? If we do, what will be energy vectors of the future? As a one-stop technology provider, Air Liquide Engineering & Construction showcases an eMethanol study for Spain, highlighting important relationships between the dimensioning of RES, electrolysis plant, Hydrogen storage and methanol synthesis. We show how Hydrogen storage and our flexible CO2-to-Methanol synthesis contribute to an integrated eMethanol plant design, suited for standalone operation with RES.

Energy Transition from Air Liquide's Perspective - What Are We Already Doing and What Lies Ahead?

Authors

Gepert, V. - Presenter

Energy transition is accelerating around the world. Besides its influence on our societies it will reshape the industry in the long-term. This is both a challenge and an enormous pool of opportunities for a better future.

One of these challenges is the fact that it is a very complex scenario with a variety of influencing factors leading to impacts in different regions and countries, e.g., the regulatory environment with specific measures related to carbon pricing, bans, social acceptance, and of course local infrastructural conditions. This increasing complexity of the industrial business environment requires adaption to do things differently.

One of the energy transition opportunities is that new technologies have to be developed and industrialized at large scale in order to address the challenge of CO2 abatement according to the Paris agreement. Decarbonizing our industry with Carbon Capture as a service (CCS/CCU) is one important lever. Key technology developments also cover low carbon H2 generation either by decarbonizing fossil fuel-based technologies or electrolysis. Increasing energy efficiency and flexibility in addition to securing ppaâ€™s is key in air gases production. Providing new solutions with regard to CO2 utilization or using bio feedstocks or wastes are additional aspects.

Hydrogen is a cornerstone energy carrier. Mastering the whole hydrogen value chain ranging from energy sourcing, production and conditioning to the end-use is key. The role of hydrogen in the energy transition is broad, as it will help to manage intermittent renewable energies and serves the areas of power generation, distribution, storage, transport, industry energy usage and raw materials. Of course access to low-carbon electricity and build-out of low-carbon power assets, CO2 storage availability and strategic partnerships are a must.

This talk shares some examples, visions and thoughts for a successful transformation towards security of supply, competitiveness and climate neutrality.

Syngas Upgrading Using a Combined Membrane-Methanation Unit: An Exhaustive Study of Critical Process Parameters on the Wobbe Index of the Injected Gas

Authors

Mirdrikvand, M. - Presenter

Jordan, U., Hitachi Zosen Inova BioMethan

The refinement of the generated raw synthesis gas in gasifiers plays a vital role in the production of bio SNG (Synthetic Natural Gas) and achieving the concept of CO2 neutral process chain. In the current process design, a practical solution for the conversion of syngas components such as CO_2 and CO to SNG with the required quality of the grid specification is introduced. The syngas undergoes a process chain of pressurized water scrubber to achieve a stochiometric relation of CO_2 and H2 to perform methanation reaction, converting CO and CO_2 to methane. In this process train, the scrubbed gas is utilized as the reactor input in a hydrogenation reactor, combining water-gas-shift and methanation reactions. To achieve appropriate spec quality of the SNG for the gas grid, a concentration of hydrogen below 0.5% vol. and a CO_2 concentration of 1% was achieved by implementing a membrane unit increasing the Wobbe index of the SNG. The optimal operational process condition for the entire process, including the scrubber, the membrane unit, and the methanation unit was achieved. The process was modeled using Aspen Custom Modeler and Aspen plus software. An in-house developed Fortran model was developed for the membrane and methanation units according to experimental data. Finally, the critical process parameters that affect the design of the membrane modules and methanation units were evaluated to achieve calorific heating value for the upgraded SNG gas.

Keywords: Syngas; methanation; membrane; retentate; permeation; methane

Ammonia Syngas Plant: Process Improvement Using Multi-Input-Multi-Output (MIMO) Surrogate Models

Authors

Malek, L. - Presenter, CGC Capital-Gain Consultants GmbH

Fricke, A.

Lukec, I., CGC Capital-Gain Consultants GmbH

Oliveira de Medeiros, D. W., Self-employed

Ammonia (NH_3) is among the most manufactured industrial chemicals in the world. In most cases, it is produced from natural gas that is taken through a series of complex reaction steps together with steam and air. The key to successful process optimization is a good understanding of how various process variables interact. The process is characterized by high energy demands, complex reaction paths, and relatively high operating temperatures and pressures.

Process interactions that are directly influencing production and energy efficiency can become increasingly complex to detect. The application of detailed process models and common sensitivity studies is often not enough for making correct optimization decisions when a process shows as complex and non-linear variable interactions as the Ammonia Syngas Plant does.

The application of an extensive Multivariate Sensitivity Study (MSS) allows the simultaneous examination of multiple input and output relationships and their effect on simulation results. MSS contributes significantly to the understanding how changes in the plant's operating conditions influence conversions and energy efficiencies and therefore increases process knowledge. The data generated by the MSS is transformed into surrogate models (SM), which are used to set optimal operating conditions to achieve economical benefits.

In this contribution, a multivariate sensitivity study is carried out on a cluster of DWSIM instances to calculate data and to train a multi-input-multi-output surrogate model. We present results from an optimization for energy efficiency by simultaneous adaptation of 9 process variables.

Process Synthesis and Optimization of Energy Integrated Distillation Sequences for Natural Gas Liquids Separation

Authors

Li, Q. - Presenter, Huazhong Agricultural University

Finn, A., Costain

Doyle, S. J., University of Manchester

Smith, R., The University of Manchester

Kiss, A. A., Delft University of Technology

Natural gas liquids (NGL) fractionation is one of the most energy-demanding process in the oil and gas industry. Due to the massive scale of operations, energy savings for the NGL fractionation are important and would contribute to less greenhouse emissions. The synthesis of energy efficient NGL distillation sequences is challenging due to many interconnected degrees of freedom associated with the large number of structural options and operating conditions. The synthesis requires to determine the basic separation configuration, including thermally coupled complex columns, as well as the column pressures, reflux ratios, feed conditions, condenser types and heat integration arrangements. For all but the simplest separation problems an exhaustive search is not feasible, and it is also not clear whether short-cut distillation models provide the necessary accuracy for reliable screening, and whether heat integration must be included simultaneously. To solve these problems, this study explores the basis on which reliable screening can be carried out. A screening algorithm has been developed using a superstructure optimization approach equipped with short cut models for the sequence synthesis, in conjunction with a transportation algorithm for the synthesis of the heat integration arrangement. Different approaches for the inclusion of heat integration are explored and compared. Then the best few NGL fractionation designs from this screening are evaluated using rigorous simulation. It has been found that for separation problems of the type explored can be screened reliably using shortcut distillation models in conjunction with the synthesis of heat exchanger network designs. Unintegrated designs using thermally coupled complex columns show much better performance than the corresponding designs using simple columns. However, once heat integration is included the difference between designs using complex columns and simple columns narrows significantly. Finally, the most techno-economically sequences for the NGL fractionation process are identified and their operating conditions are detail designed.

Determining DSC Heat Release Rate Sensitivity to Improve Thermal Safety Evaluation

Authors

Lousa, N. - Presenter, Hovione Farmaciencia, S.A.

Gaspar De Sousa, L., Hovione Farmaciencia SA

Alvarez, C., Hovione Farmaciencia SA

For more than two decades, different methods for estimating the time to maximum rate under adiabatic conditions (TMR_{ad}) have been developed. When it comes to the traditional quick screening using non-adiabatic experiments, Keller et al and Stoessel work have set the industry gold-standard by using a single dynamic DSC run with milligrams of material. Advances in thermal analysis and computational power gave rise to isoconversional methods, where a more complex and lengthy DSC experimental protocol is required but return much accurate and user-defined scenario outputs when compared to the traditional approach. In the work described herein, we leveraged from the knowledge obtained in DSC analysis done for thermal safety evaluation purposes and were able to estimate the heat release rate at T_{onset} for any material handled at Hovione plants after a single DSC run. This enables the calculation of a TMR_{ad} with a better compromise between speed and the intrinsic conservative nature of the previous traditional safety calculations, thus improving risk perception and avoiding unnecessary red flags or time-consuming actions during process chemistry development.

End-to-end bioprocessing models to assess process robustness and inform control strategies

Authors

Birmingham, S. - Presenter, Siemens

Close, E., Siemens

In this talk we will describe examples of how Siemens customers are building and applying dynamic end-to-end mechanistic models of integrated and continuous biomanufacturing processes (ICB) to determine the impact of expected disturbances, deviations, and uncertainties on product quality. We will show practical examples where these models can generate value by performing tasks which are not possible experimentally due to the prohibitive material requirements and complexity of building end-to-end processes in the lab.

Controllability of cell population in bioreactor is mainly determined by the fitness cost associated with phenotypic switching

Authors

Delvigne, F. - Presenter, University of Liège

Lucas Henrion[1], Juan Andres Martinez[1], Vincent Vandenbroucke[1], Mathéo Delvenne[1], Samuel Telek[1], Andrew Zicler[1], Alexander Grünberger[2], Frank Delvigne[1*]

[1]Terra research and teaching centre, Microbial Processes and Interactions (MiPI), Gembloux Agro-Bio Tech, University of Liège, Gembloux, Belgium

[2]Microsystems in Bioprocess Engineering, Institute of Process Engineering in Life Sciences, Karlsruhe Institute of Technology, Karlsruhe, Germany

*Corresponding author: F.Delvigne@uliege.be

Clonal population of cells can cope with stress conditions by switching to alternative phenotypes. Even if it can lead to increased fitness in a natural context, this feature is typically unwanted for a range of applications (e.g., bioproduction, synthetic biology, biomedicine...) where it tends to decrease the controllability of the cellular response. However, little is known about the diversification profiles that can be adopted by a cell population. We characterized the diversification dynamics for 6 different biological systems involving Gram-negative and Gram-positive bacteria, as well as eukaryotes and for different phenotypes (utilization of alternative carbon sources, general stress response and more complex development patterns). Interestingly, our results suggest that the diversification dynamics and the fitness cost associated with cell switching are coupled. Furthermore, we used a cell-machine interface (Segregostat) to demonstrate that the phenotypes associated with these diversification regimes exhibit different levels of controllability, enabling applications involving more precise cellular responses.

Keywords: phenotypic switching, population control, automated flow cytometry, biological noise, bet hedging, information theory

Analyzing bioprocess heterogeneity from the microbial viewpoint: recent developments

Authors

Haringa, C. - Presenter

There is considerable in the impact of environmental- and population heterogeneity on industrial bioprocess. One of the tools that has gained popularity is so-called â€˜lifeline analysisâ€™, where temporal fluctuations in the environment observed by a cell are analyzed. Analysis of such lifelines can provide insight in environmental stresses affecting the microbial response and thereby process performance, and forms a basis for bioreactor scale-down [1].

The most common form of lifeline analysis is done through Euler-Lagrange Computational Fluid Dynamics (CFD). A main bottlenecks for routine application of lifeline concepts is the high computational expense, but recent advances are alleviating this bottleneck. In particular, GPU-based Lattice-Boltzmann methods are promising: more accurate results are acquired with reduced computation time. We have utilized GPU-driven Lattice-Boltzmann simulations for parcel-based downscaling studies [2], and more recently, two-way coupling of structured biokinetics has been introduced in the Lattice-Boltzmann framework. These advances allow for much faster evaluation of performance snapshots of bioreactors, but CFD remains too computationally expensive to analyze full bioprocesses. To this end, stochastic particle tracking in compartment models with structured kinetics forms an interesting route. [3]

Besides this computational lifeline analysis, developments regarding flow-following sensor particles and the use of microfluidic single-cell cultivations to mimic lifelines experimentally open the door to a more interdisciplinary concept [4]. I will briefly discuss this broader concept, and provide a perspective how lifeline analysis can contribute to bioprocess evaluation, optimization and operation.

[1]: Haringa et al., Biochemical Engineering Journal (2018)

[2]: Haringa, Engineering in Life Sciences (2022)

[3]: Haringa, Tang, Noorman, Biotechnology & Bioengineering (2022)

[4]: Blöbaum, Haringa, Grünberger, Biotechnology Advances (2023)

Cybergenetics in the Bioreactor: Optimizing Biprocesses through Light-Based Feedback Control

Authors

Khammash, M. - Presenter, ETH Zürich

Cybergenetics is an emerging field concerned with the robust closed-loop control of biological processes at the genetic level. In this talk, I will present our approach of combining synthetic biology, optogenetics, and control engineering to enable cybergnetic control of biotechnological production processes. After describing the technological setups needed for this approach, I will demonstrate the application of cybergenetics in two different applications: control of bacterial co-culture composition, and the control of the unfolded protein response in S. cerevisiae for improved protein production.

Importance of (dynamic) Process Simulation on Digital Transformation and Net Zero Emissions

Authors

Mousazadeh, F. - Presenter, Delft University of Technology

Digital transformation (specially process simulation as essential element) and net zero emission are two topics which are strongly related. Digital transformation is a journey which is time consuming as well as unclear for organizations how to proceed. Process Technology has a lot of uncertainties (reliability of new technologies, added value of the transformation, transformation roadmap, etc) before starting the digital transformation as well as during the transformation.

To sustainable transformation journey and towards net zero emissions, accurate modelling of existing technology and the new design concepts will help in understanding the performance of these options and assist in taking decisions. Confident deployment of capital, allocation of limited financial and human resources, and efficient execution of complex projects for restructuring and re-alignment are equally important. By using process simulation it will be possible to map interconnected processes and energy systems down to the heater, boiler, compressor and any other relevant process equipment and it allows the emissions to be quantified, prioritized, monitored, and then eliminated using a range of solutions such as electrification, hydrogen firing, reduced flaring, etc..

In addition, a process digital twin, with a process simulation/production model at its core, is a powerful tool to help design, monitor, troubleshoot, optimize, forecast, and plan single unit and full asset performance. It provides underlying performance information, automates the collection and validation of data, and delivers the analysis through visualization and tailored reporting.

Modularization of Large Process Flowsheeting Models

Authors

Mardani, S. - Presenter

Al haj Ali, M. - Presenter, Borealis Polymers Oy

Conwell, R., Borealis

The concept of digital twin(DT) as a virtual model of a physical process was originally proposed by Grieves[1]. DTs are aiming to simulate a process in a digital environment and compare it with the physical plant, targeting various benefits like process optimization and anomaly detection[1].

Chemical plants consist of process equipment and instrumentation, making the processes very complex[2]. Consequently, the corresponding plant-scale DTs are equally complex, resulting in three major issues. First, the simulation *speed* (real-time factor) of high-fidelity, plant-scale dynamic models is low. Second, the *scope* of plant-scale simulations creates operational challenges for the users. Third, the *versatility* of the tool is limited, when exploring scenarios that require changes in the process model, since this exercise is often time consuming and requires specific expertise.

To overcome these issues, we proposed a *modular* approach, where the main flowsheet (covering the full asset), is composed of sub-flowsheets, modelling the different areas of the plant (i. e. compressor area, reactor area, etc.). Within each sub-flowsheet, it is possible to design several modules, where each module shares the same inputs / outputs. However, the internal design of the each module has a different level of fidelity. For example, a compressor area sub-flowsheet might contain a full-fidelity module and a low-fidelity module. The full-fidelity module uses a rigorous centrifugal compressor model, including the ancillary equipment items. The low-fidelity module consists of a simplified compressor with a specified discharge pressure. The simulation operator activates either, the full-fidelity, or the low-fidelity module, based on the fidelity needed by his research objectives. This modularization provides a *versatile* selection of process *scope*, increasing the processing *speed* of the targeted research.

1. Michael Grieves, white paper (2014), Michael W. Grieves, LLC
2. Eng. Res. Des. 87 (2009) 1430â€"1437

Python in Chemical Process Simulation

Authors

Fricke, A. - Presenter

Soni, J., CGC Capital-Gain Consultants GmbH

Malek, L., CGC Capital-Gain Consultants GmbH

Oliveira de Medeiros, D. W., Self-Employed

Python is a powerful programming language that has become increasingly important in the field of chemical engineering. One reason for its growing popularity is its versatility and ability to handle a wide range of tasks. It offers libraries, such as SciPy and NumPy to perform complex mathematical calculations, or libraries for data analysis, such as pandas and scikit-learn to work with large sets of data.

Despite its many strengths, Python is not currently able to fully replace specialized process simulation software in chemical engineering. These simulation software packages, such as Aspen Plus, HYSYS, CHEMCAD, ProII and DWSIM, have been developed specifically for use in the chemical industry and have many more advanced features than those available in Python libraries. For example, they may have more detailed thermodynamic models, a more comprehensive library of chemical compounds, or pre-built unit operations, e.g., distillation columns and reactors.

There are several ways to integrate Python with a chemical process simulator depending on the specific chemical process simulator being used and the desired level of integration. Directly integrating Python with a chemical process simulator by using its Application Programming Interface (API) allows to run the chemical process simulator and automate it with Python scripts.

This enables to combine simulator-internal and external optimizers creating complex simulations that can be used to analyze and predict chemical processes, which aids in the design and optimization of chemical processes, leading to more efficient and sustainable, and cost-effective production.

This contribution shows how HYSYS, CHEMCAD, and DWSIM can be used from Python to create surrogate models (SM) from flowsheet simulations (PFD), embed SM in PFDs, optimize PFDs using SM, and create new PFDs using AI assistance.

CO_2 Freezing in the Low-Temperature Separation of Carbon Dioxide from Natural Gas- By Distillation

Authors

Adnan, N. - Presenter

Smith, R., The University of Manchester

Zhang, N., University of Manchester

Jobson, M., The University of Manchester

Natural gas is a cleaner alternative to coal and oil. This increases the demand for natural gas and creates an incentive to exploit CO_2-rich gas reserves. Extracted gas is increasingly transported as liquefied natural gas (LNG) over long distances, which is more cost-effective than pipelines. This study explores solutions to remove CO_2 using low-temperature distillation to meet the 50 ppm requirement for liquefaction of CO_2-rich natural gas reserves. Distillation can produce high-purity natural gas and liquid CO_2 suitable for sequestration but low temperatures can cause CO_2 to freeze, creating operational difficulties for the distillation.

This study investigates the freezing behaviour of CO_2 in distillation operations considering a mixture of methane and impurities to understand the range of conditions and compositions where CO_2 freezing occurs. Phase equilibrium predictions using Aspen HYSYS are verified against the published experimental data available and a published thermodynamic model.

Simulations of the process consisting of a sequence of two distillation columns considering binary, ternary and multicomponent mixtures have been carried out in Aspen HYSYS, using the CO_2 freeze-out tool to predict freezing. This study also explores approaches to avoid CO_2 freezing in both columns including varying pressure and intermediate CO_2 mol fraction to alter freezing temperatures.

This study finds that the condenser of the first column and the stripping section of the second column are most vulnerable to freezing. The presence of nitrogen exacerbates freezing in the second column, while the presence of heavier hydrocarbons lowers the CO_2 freezing temperature. When considering a natural gas feed system, the simulations found that CO_2 freezing in the first column can be avoided by specifying the process at high pressures and high intermediate CO_2 mole fraction in the distillate.

Future work aims to adopt optimal design and operation of the distillation process with process intensification for offshore applications.

On the Revival of a 500-Year-Old Near-Extinct Artisanal Salt-Making: The Asin Tibuok Process and Chemical Engineering Integration Towards Sustainable Production

Authors

Javier, J. O. - Presenter, University of the Philippines Diliman

Macay, K. I. - Presenter, Department of Chemical Engineering, University of the Philippines Diliman, Quezon City, Philippines, 1101

Salvador, M. A. - Presenter, Department of Chemical Engineering, University of the Philippines Diliman, Quezon City, Philippines, 1101

Singh, M. - Presenter, Department of Chemical Engineering, University of the Philippines Diliman, Quezon City, Philippines, 1101

Lerona, P. G. - Presenter, Department of Chemical Engineering, University of the Philippines Diliman, Quezon City, Philippines, 1101

Dela Cruz, I. - Presenter, Department of Chemical Engineering, University of the Philippines Diliman, Quezon City, Philippines, 1101

Asin Tibuok, which literally means â€œunbroken saltâ€� or â€œwhole saltâ€�, is a Bohol-native gourmet salt that resembles that of a dinosaur egg and has been recently gaining international traction. The process of producing Asin Tibuok, an artisanal salt, is delicate because of the nature of how it is traditionally and meticulously made. It uses eleven (11) steps that take up to 4 months to finish. This is one of the main reasons why new generations of the *asinderos* have lost interest in the tradition of Asin Tibuok production. This declining interest was worsened during the onslaught of supertyphoon Odette (International name: Typhoon Rai) which damaged most of the facility. Hence, it was imperative to document the traditional and cultural process while also incorporating chemical engineering methods that ensure the continuous production of the salt. This work is comprised of a market study, unit process design, and economic analysis. Through the market study, necessary information on global and local supply and demand of product and raw materials were obtained and reviewed to survey a sustainable market for Asin Tibuok. Equipment sizing and its corresponding streams were modeled to design a plant which has a capacity of 200 units per batch. Lastly, subsequent economic analysis of the optimized design suggests a sustainable and feasible production while preserving the cultural heritage of the Asin Tibuok production.

Optimization of High-Density Polyethylene (HDPE) Production Via Ziegler-Natta-Catalyzed Slurry-Phase Polymerization Process for the Philippine Chemical Industry

Authors

Lee, P. S. - Presenter, University of the Philippines Diliman

Esguerra, A. M. - Presenter, University of the Philippines Diliman

Labongray, H. D. - Presenter, University of the Philippines Diliman

Lerona, P. G. - Presenter, Department of Chemical Engineering

Slurry phase polymerization is a well-established process in polymer technology that has been used for decades to produce a wide range of polymers, including polyethylene and polypropylene. It involves the polymerization of monomers in a liquid medium, usually a solvent or a mixture of solvents, to form a suspension of polymer particles. However, this process has not been explored and implemented by polyethylene manufacturers in the Philippines. Moreover, recent studies using slurry-phase polymerization models have exhibited low monomer conversion. Hence, this work performed a series of optimization studies to examine the effects and determine the optimum reactor and feed conditions. Reactor configuration was also optimized using sensitivity analysis with consistent polymer attributes. Using Perturbed-Chain Statistical Associating Fluid Theory (PC-SAFT) and Ziegler-Natta polymerization models, the incorporation of optimization and balance blocks led to convergent results and was proven to increase monomer conversion. The optimized model developed in this study has the potential to increase HDPE production in the country.

Membrane Contactor Applied for Acrylic Acid Removal from Acrylate Downstream Processing: An Environmental Techno-Economic Assessment

Authors

Yang, J. - Presenter

In acrylates industry, the conventional acrylic acid (AA) removal process using a water washing tower is costly by consuming a large amount of water. To replace the traditional water washing tower for acrylates production, a TiO_2 ceramic membrane contactor to extract AA is studied. An environmental techno-economic assessment (ETEA) is performed for the acrylate production process with a membrane contactor, and its results are benchmarked with those of the traditional process in this study. A model acrylate, 2-ethylhexyl acrylate (2-EHA) is taken as the product in the assessment. Our analysis is supported by novel experimental evidence, where we found a mass transfer coefficient of acrylic acid (0.003 m^3/m^2h) as the speed at which the membrane allows transferring acrylic acid from an acrylate esterification solution. We consider the full production process from 2-EHA esterification to purification, with a higher resolution for the downstream processing. A manufacturing scale of 20,000 ton/year is assumed. We find that the membrane process has a slightly higher manufacturing cost (â‚¬1,643/ton 2-EHA) compared to the water washing process (â‚¬1,629/ton 2-EHA), due to the high membrane cost of â‚¬25/ton 2-EHA. However, the membrane process has â‚¬20/ton 2-EHA lower cost from energy consumption and other operating cost, because 30% less water is demanded by the membrane contactor. The lower water consumption also results in a lower environmental impact than the water washing process, as indicated by scores of the 18 midpoint categories. The manufacturing cost of the membrane process will be lower than the water washing process when increasing the mass transfer coefficient to 0.005 m^3/m^2h.

Optimization of Pyrolysis-Separation Process for the Production of Crude Oil from Valorization of Waste Face Masks (WFMs) from Metro Manila, Philippines

Authors

Lerona, P. G. - Presenter, Department of Chemical Engineering

Battad, B. A. - Presenter

Aguda, C. J. R. - Presenter

Liclican, A. J. - Presenter, University of the Philippines

In the Philippines, the face mask used in 2020 to prevent the spread of the COVID-19 virus led to a 595% rise in plastic-based medical waste. Current medical waste treatment methods in the country, such as landfilling, chemical disinfection, burial, and vault confinement, either harm the environment or yield no value post-processing. Repurposing is necessary to reduce greenhouse gas emissions and avoid accumulation. In this study, a proposed pyrolysis-separation process utilizes 1 ton per day of waste face masks (WFMs) obtained from nearby hospitals and medical facilities in the Metro Manila area, which is mainly composed of polypropylene (PP) and low-density polyethylene (LDPE). WFM's undergo pretreatment in a hammermill for size reduction. The hammermill fines are fed into a pyrolysis reactor producing vaporized hydrocarbons and solid char. The vapor stream is divided into non-condensable hydrocarbon gases and heavy condensable crude oil through a flash separator. The process was simulated through Aspen HYSYS using two continuous stirred tank reactors (CSTRs), and one separator in a parallel configuration, with PP and LDPE feeds in each configuration. The crude oil product, composed mainly of heavy hydrocarbons, was consistent with commercial standards. Optimization of the proposed pyrolysis-separation process led to the production of 0.97 tons per day crude oil. By utilizing WFMs, the plant is capable of reducing the waste generation of Metro Manila by 330 tons annually.

AUTHOR INDEX